T0257351

Getting Started with RFID

Tom Igoe

O'REILLY®

Beijing · Cambridge · Farnham · Köln · Sebastopol · Tokyo

Getting Started with RFID
by Tom Igoe

Published by O'Reilly Media, Inc., 1005 Gravenstein Highway North, Sebastopol, CA 95472.

O'Reilly books may be purchased for educational, business, or sales promotional use. Online editions are also available for most titles (*http://my.safaribooksonline.com*). For more information, contact our corporate/institutional sales department: (800) 998-9938 or *corporate@oreilly.com*.

Editor: Brian Jepson
Production Editor: Teresa Elsey
Cover Designer: Mark Paglietti
Interior Designers: Ron Bilodeau and Edie Freedman
Illustrators: Robert Romano and Rebecca Demarest

March 2012:   First Edition.

Revision History for the First Edition:
    March 09, 2012   First release
See *http://oreilly.com/catalog/errata.csp?isbn=9781449324186* for release details.

The font used in the figures is Architects Daughter, provided by Google Web Fonts under SIL Open Font License 1.1.

ISBN: 978-1-449-32418-6
[LSI]
1331238342

Contents

Preface

The process of identifying physical objects is such a fundamental part of our experience that we seldom think about how we do it. We use our senses, of course: we look at, feel, pick up, shake and listen to, smell, and taste objects until we have a reference for them—then we give them a label. The whole process relies on some pretty sophisticated work by our brains and bodies, and anyone who's ever dabbled in computer vision or artificial intelligence in general can tell you that teaching a computer to recognize physical objects is no small feat. Just as it's easier to determine location by having a human being narrow it down for you, it's easier to distinguish objects computationally if you can limit the field, and if you can label the important objects.

Just as we identify things using information from our senses, so do computers. They can identify physical objects only by using information from their sensors. One of the best-known digital identification techniques is *radio frequency identification*, or RFID. The network identity of a physical object can be centrally assigned and universally available, or it can be provisional. It can be used only by a small subset of devices on a larger network or used only for a short time. RFID is an interesting case in point. The RFID tag pasted on the side of a book may seem like a universal marker, but what it means depends on who reads it. The owner of a store may assign that tag's number a place in his inventory, but to the consumer who buys it, it means nothing unless she has a tool to read it and a database in which to categorize it. She has no way of knowing what the number meant to the store owner unless she has access to his database. Perhaps he linked that ID tag number to the book's title or to the date on which it arrived in the store. Once it leaves the store, he may delete it from his database, so it loses all meaning to him. The consumer, on the other hand, may link it to entirely different data in her own database, or she may choose to ignore it entirely, relying on other means to identify it. In other words, there is no central database linking RFID tags and the things they're attached to, or who's possessed them.

Like locations, identities become more uniquely descriptive as the context they describe becomes larger. For example, knowing that my name is Tom doesn't give you much to go on. Knowing my last name narrows it down some more, but how effective that is depends on where you're looking. In the United States, there are dozens of Tom Igoes. In New York, there are at least three. When you need a unique identifier, you might choose a universal label, like using my Social Security number, or you might choose a provisional label, like calling me "Frank's son Tom." Which you choose depends on your needs in a given situation. Likewise, you may choose to identify physical objects on a network using universal identifiers, or you might choose to use provisional labels in a given temporary situation.

The capabilities assigned to an identifier can be fluid as well. Taking the RFID example again: in the store, a given tag's number might be enough to set off alarms at the entrance gates or to cause a cash register to add a price to your total purchase. In another store, that same tag might be assigned no capabilities at all, even if it's using the same protocol as other tags in the store. Confusion can set in when different contexts use similar identifiers. Have you ever left a store with a purchase and tripped the alarm, only to be waved on by the clerk who forgot to deactivate the tag on your purchase? Try walking into a Barnes & Noble bookstore with jeans you just bought at a Gap store, and you might trip the alarms if the two companies use the same RFID tags but don't set their security systems to filter out tags that are not in their inventory.

 NOTE: This short book presents a couple of RFID projects for Processing and Arduino from the first edition of *Making Things Talk* (O'Reilly 2007). When this book was updated to a second edition in 2011, the RFID examples were updated to work with newer RFID readers, specifically those that interoperate with the Near-Field Communications (NFC) readers found in mobile phones such as the Nexus S. Because there is still interest in the Parallax RFID reader used in the first edition, this book is here to preserve those projects for anyone who's interested in building them.

Who This Book Is For

If you've got some experience with Arduino and Processing, and are curious to experiment with radio frequency identification, this book is for you. You won't need any advanced skills: as long as you know enough about Arduino and Processing to run simple sketches, and are able to connect basic circuits on a breadboard with jumper wire, you'll be able to use this book. If you don't have any experience with Arduino or Processing, the book *Getting Started with Arduino*, second edition, by Massimo Banzi (O'Reilly) and *Getting*

Started with Processing by Casey Reas and Ben Fry (O'Reilly) will get you started.

Companion Kit

A kit is coming soon from Maker Shed to go along with this book. It will include all the components you'll need, from the Arduino to the RFID reader. For more information, see *http://www.makershed.com/*.

Conventions Used in This Book

The following typographical conventions are used in this book:

Italic
> Indicates new terms, URLs, email addresses, filenames, and file extensions.

`Constant width`
> Used for program listings, as well as within paragraphs to refer to program elements such as variable or function names, databases, data types, environment variables, statements, and keywords.

`Constant width bold`
> Shows commands or other text that should be typed literally by the user.

`Constant width italic`
> Shows text that should be replaced with user-supplied values or by values determined by context.

 TIP: This icon signifies a tip, suggestion, or general note.

 CAUTION: This icon indicates a warning or caution.

Using Code Examples

This book is here to help you get your job done. In general, you may use the code in this book in your programs and documentation. You do not need to contact us for permission unless you're reproducing a significant portion of the code. For example, writing a program that uses several chunks of code from this book does not require permission. Selling or distributing a CD-ROM of examples from O'Reilly books does require permission. Answering a question by citing this book and quoting example code does not require permission. Incorporating a significant amount of example code from this book into your product's documentation does require permission.

We appreciate, but do not require, attribution. An attribution usually includes the title, author, publisher, and ISBN. For example: "*Getting Started with RFID* by Tom Igoe (O'Reilly). Copyright 2012 Tom Igoe, 978-1-449-32418-6."

If you feel your use of code examples falls outside fair use or the permission given above, feel free to contact us at *permissions@oreilly.com*.

Safari® Books Online

 Safari Books Online (*www.safaribooksonline.com*) is an on-demand digital library that delivers expert content in both book and video form from the world's leading authors in technology and business. Technology professionals, software developers, web designers, and business and creative professionals use Safari Books Online as their primary resource for research, problem solving, learning, and certification training.

Safari Books Online offers a range of product mixes and pricing programs for organizations, government agencies, and individuals. Subscribers have access to thousands of books, training videos, and prepublication manuscripts in one fully searchable database from publishers like O'Reilly Media, Prentice Hall Professional, Addison-Wesley Professional, Microsoft Press, Sams, Que, Peachpit Press, Focal Press, Cisco Press, John Wiley & Sons, Syngress, Morgan Kaufmann, IBM Redbooks, Packt, Adobe Press, FT Press, Apress, Manning, New Riders, McGraw-Hill, Jones & Bartlett, Course Technology, and dozens more. For more information about Safari Books Online, please visit us online.

How to Contact Us

Please address comments and questions concerning this book to the publisher:

O'Reilly Media, Inc.
1005 Gravenstein Highway North
Sebastopol, CA 95472
800-998-9938 (in the United States or Canada)
707-829-0515 (international or local)

707-829-0104 (fax)

We have a web page for this book, where we list errata, examples, and any additional information. You can access this page at:

http://shop.oreilly.com/product/0636920024842.do

To comment or ask technical questions about this book, send email to:

bookquestions@oreilly.com

For more information about our books, courses, conferences, and news, see our website at *http://www.oreilly.com*.

Find us on Facebook: *http://facebook.com/oreilly*

Follow us on Twitter: *http://twitter.com/oreillymedia*

Watch us on YouTube: *http://www.youtube.com/oreillymedia*

1/Radio Frequency Identification

Like bar code recognition, RFID relies on tagging objects in order to identify them. Unlike bar codes, however, RFID tags don't need to be visible to be read. An RFID reader sends out a short-range radio signal, which is picked up by an RFID tag. The tag then transmits back a short string of data. Depending on the size and sensitivity of the reader's antenna and the strength of the transmission, the tag can be several feet away from the reader, enclosed in a book, box, or item of clothing. In fact, some large clothing manufacturers are now sewing RFID tags into their merchandise, to be removed by the customer.

There are two types of RFID system: passive and active. Passive RFID tags contain an integrated circuit that has a basic radio transceiver and a small amount of nonvolatile memory. They are powered by the current that the reader's signal induces in their antennas. The received energy is just enough to power the tag to transmit its data once, and the signal is relatively weak. Most passive readers can only read tags a few inches to a few feet away.

In an active RFID system, the tag has its own power supply and radio transceiver, and transmits a signal in response to a received message from a reader. Active systems can transmit for a much longer range than passive systems, and are less error-prone. They are also much more expensive. If you're a regular automobile commuter and you have to pass through a toll gate in your commute, you're probably an active RFID user. Systems like E-ZPass, shown in Figure 1-1, use active RFID tags so that the reader can be placed several meters away from the tag.

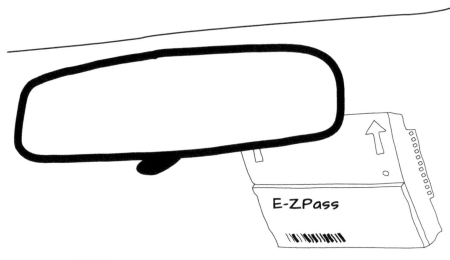

Figure 1-1. *An E-ZPass active RFID tag mounted on a car's windshield*

You might think that because RFID is radio-based, you could use it to do radio distance ranging as well, but that's not the case. Neither passive nor active RFID systems are typically designed to report the signal strength received from the tag. Without this information, it's impossible to use RFID systems to determine the actual location of a tag. All the reader can tell you is that the tag is within reading range. Although some high-end systems can report the tag signal strength, the vast majority of readers are not made for location as well as identification. You can do some limited location detection using multiple readers, however. Because most passive RFID readers (like the one used in this book) have a short range, you can be assured that if you get a signal from a tag on a particular reader, the tag is within a few centimeters of the reader. By using an array of readers spaced half a meter or so apart, you could determine the rough location of a tag by knowing that it's in a given reader's range.

RFID systems vary widely in cost. Active systems can cost tens of thousands of dollars to purchase and install. Commercial passive systems can also be expensive. A typical passive reader that can read a tag a meter away from the antenna typically costs a few thousand dollars. At the low end, short-range passive readers can come as cheap as $30 or less. As of this writing, $30 to $100 gets you a reader that can read a few centimeters. Anything that can read a longer distance will be more expensive.

There are many different RFID protocols, just as with bar codes. Short-range passive readers come in at least three common frequencies: two low-frequency bands at 125 and 134.2 Khz, and high-frequency readers at

13.56MHz. The higher-frequency readers allow for faster read rates and longer-range reading distances. In addition to different frequencies, there are also different protocols. For example, in the 13.56 band alone, there are the ISO 15693 and ISO 14443 and 14443-A standards; within the ISO 15693 standard, there are different implementations by different manufacturers: Philips' I-Code, Texas Instruments' Tag-IT HF, Picotag, and implementations by Infineon, STMicroelectronics, and others. Within the ISO 14443 standard, there's Philips' Mifare, Mifare UL, ST's SR176, and others. So you can't count on one reader to read every tag. You can't even count on one reader to read all the tags in a given frequency range. You have to match the tag to the reader.

There are a number of inexpensive and easy-to-use readers on the market now, covering the range of passive RFID frequencies and protocols. Maker Shed sells a 125KHz reader (*http://www.makershed.com/product_p/ mkpx2.htm*) from Parallax that can read EM Microelectronic tags, such as EM4001 tags. It has a built-in antenna, and the whole module is about 2.5″ × 3.5″, on a flat circuit board. The EM4001 protocol isn't as common in everyday applications as the Mifare protocol, a variation on the ISO 14443 standard in the 13.56 MHz range. This book doesn't cover the Mifare tags, but *Making Things Talk*, second edition (O'Reilly, 2011) does.

RFID tags come in a number of different forms, as shown in Figure 1-2: sticker tags, coin discs, key fobs, credit cards, playing cards, even capsules designed for injection under the skin. The last are used for pet tracking and are not designed for human use, though there are some adventurous hackers who have had these tags inserted under their own skin. Like any radio signal, RFID can be read through a number of materials, but it is blocked by any kind of RF shielding, like wire mesh, conductive fabric lamé, metal foil, or adamantium skeletons. This feature means that you can embed it in all kinds of projects, as long as your reader has the signal strength to penetrate.

Before picking a reader, think about the environment in which you plan to deploy it, and how that affects both the tags and the reading. Will the environment have a lot of RF noise? In what range? Consider a reader outside that range. Will you need a relatively long-range read? If so, look at the active readers, if possible. If you're planning to read existing tags rather than tags you purchase yourself, research carefully in advance, because not all readers will read all tags. Pet tags can be some of the trickiest, as many of them operate in the 134.2 KHz range, where there are fewer readers to choose from.

In picking a reader, you also have to consider how it behaves when tags are in range. For example, even though the Parallax reader used in this book and a compatible reader from ID Innovations can read the same tags, they be-

Figure 1-2. *Various RFID tags*

have very differently when a tag is in range. The ID Innovations reader reports the tag ID only once. The Parallax reader reports it continually until the tag is out of range. The behavior of the reader can affect your project design, as you'll see later on.

The Parallax reader has a TTL serial interface, so it can be connected to a microcontroller or a USB-to-serial module very easily. You'll see sketches in Processing (which run on a computer and connect to the reader over USB-to-serial) for the Parallax reader in this book, in fact.

2/Reading RFID Tags in Processing

In this project, you'll read some RFID tags and get a sense of how the readers behave. You'll see how far away from your reader a tag can be read. This is a handy test program for use any time you're adding RFID to a project.

Materials

RFID reader
> Parallax's RFID Reader Module, available from Maker Shed as part of a starter pack (*http://www.makershed.com/product_p/mkpx16.htm*) or by itself (*http://www.makershed.com/product_p/mkpx2.htm*).

RFID tags
> The starter pack (*http://www.makershed.com/product_p/mkpx16 .htm*) includes several tags, and you can buy them (*http://www.make rshed.com/SearchResults.asp?Search= rfid*)
> separately (*http://www.makershed.com/SearchResults.asp?Search=rfid*).

USB-to-TTL serial adaptor
> The FTDI Friend (*http://www.makershed.com/FTDI_Friend_v1_0_p/ mkad22.htm)can*) can do the job.

Breadboard
> You can use a half-size (*http://www.makershed.com/product_p/ mkkn2.htm*) or mini breadboard (*http://www.makershed.com/Mini _Breadboards_p/mkkn1.htm*) to make connections between the reader and the USB-to-TTL serial adaptor.

Jumper wire
> You'll need a set of jumper wire (*http://www.makershed.com/product _p/mkseeed3.htm*) to make your connections.

Parallax RFID Reader

The Parallax reader is one the simplest readers available. It communicates serially at 2400 bps. When the Enable pin is held low (connected to ground), it sends a reading whenever a tag is in range. The tag ID is a 12-byte string starting with a carriage return (ASCII 13) and finishing with a newline (ASCII 10). The ten digits in the middle are the unique tag ID. The EM4001 tags format their tag IDs as ASCII-encoded hexadecimal values, so the string will never contain anything but the ASCII digits 0 through 9 and the letters A through F.

The Circuit

The circuit for this reader is very simple. Connect the module to 5V and ground, and connect the reader's serial transmit line (labeled SOUT) to the serial adaptor's serial receive line (labeled RX). You'll also need to attach the enable pin to ground. Figure 2-1 shows these connections.

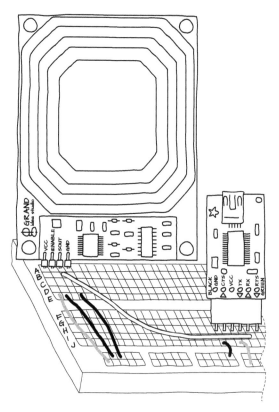

Figure 2-1. *Wiring the RFID reader to the FTDI Friend*

Try It

The following Processing (*http://processing.org/*) sketch waits for twelve serial bytes, strips out the carriage return and the newline, and prints the rest to the screen. Before you run this sketch, plug the FTDI Friend into your computer with a USB Mini cable.

 NOTE: You will probably need to look at the output of `Serial.list()` and change the number on the next line of code match the serial port that corresponds to your microcontroller.

```
/*
 Parallax RFID Reader
 language: Processing

 */

// import the serial library ❶
import processing.serial.*;

Serial rfidPort;       // the serial port you're using
String tagID = "";   // the string for the tag ID

void setup() {
  size(600, 200);
  // list all the serial ports ❷
  println(Serial.list());

  // based on the list of serial ports printed from the
  // previous command, change the 0 to your port's number ❸
  String portnum = Serial.list()[0];

  // initialize the serial port ❹
  rfidPort = new Serial(this, portnum, 2400);

  // incoming string from reader will have 12 bytes:
  rfidPort.buffer(12);

  // create a font with the third font available to the system:
  PFont myFont = createFont(PFont.list()[2], 24);
  textFont(myFont);
}
```

```
void draw() {
  // clear the screen:
  background(0);

  // print the string to the screen ❺
  text(tagID, width/4, height/2 - 24);
}

/*
 this method reads bytes from the serial port
 and puts them into the tag string.
 It trims off the \r and \n
*/
void serialEvent(Serial rfidPort) { // ❻
  tagID = trim(rfidPort.readString());
}
```

Here's an explanation of the key parts of the code:

❶ This line imports the serial library that comes with Processing. With it, you'll be able to use serial functions later in this sketch.

❷ This line will print all the available serial ports to the Processing console. You should examine the output of this command and identify which of your serial ports corresponds to the FTDI Friend that's plugged into your computer.

❸ If the output of the previous line of code was anything other than the first serial port (index 0), change the 0 in this line to the index of the correct serial port.

❹ This line begins serial communications at 2400 bits per second.

❺ Processing's `draw()` function runs continuously as long as the sketch is running. This line will display whatever's in the `tagID` variable to the screen, even if it's still blank.

❻ This function is invoked any time there's some incoming activity on the serial port. When the Parallax reader sends something to the Processing sketch, it will be a tag id. This line puts that tag's ID into the `tagID` variable.

Figure 2-2 shows the results of holding a tag up to the reader while this Processing sketch is running.

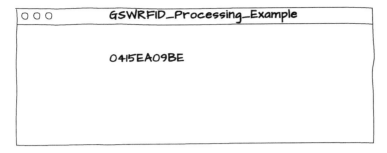

Figure 2-2. *The Processing sketch reading a tag*

3/Reading RFID Tags in Arduino

In this project, you'll connect Arduino directly to the RFID reader. This project accomplishes the same thing as the project in Chapter 2, but the reader is connected to an Arduino, not to your computer. As a starter step, you'll see how to read in an RFID tag's ID and send its value to a computer over the serial port. After you've done that, you'll see how to read in an RFID tag's ID and compare it to a stored tag ID: if you wave the right tag at the RFID reader, it will light an LED. In this way, the RFID tag will behave as a key.

WARNING: Some RFID tags have many well-documented vulnerabilities. Certain types of RFID tags can be cloned easily, for example. So in theory, if an attacker gets close enough to your RFID tag (or you code) to determine the ID of the tag, they may be able to create a copy of it.

Materials

Arduino
   The Arduino Uno (*http://www.makershed.com/product_p/mksp4.htm*) is a good model of Arduino to get started with for all the micro-controller-based projects in this book.

RFID reader
   Parallax's RFID Reader Module, available from Maker Shed as part of a starter pack (*http://www.makershed.com/product_p/mkpx16.htm*) or by itself (*http://www.makershed.com/product_p/mkpx2.htm*).

RFID tags
   The starter pack (*http://www.makershed.com/product_p/mkpx16.htm*) includes several tags, and you can buy them separately (*http://www.makershed.com/SearchResults.asp?Search=rfid*).

Breadboard

You can use a half-size (*http://www.makershed.com/product_p/ mkkn2.htm*) or mini breadboard (*http://www.makershed.com/Mini _Breadboards_p/mkkn1.htm*) to make connections between the reader and the USB-to-TTL serial adaptor.

Jumper wire

You'll need a set of jumper wire (*http://www.makershed.com/product _p/mkseeed3.htm*) to make your connections.

The Circuit

The circuit for this reader is similar to the one from Chapter 2. Connect the module to the Arduino's 5V and ground connections, and connect the reader's serial transmit line (labeled SOUT) to digital pin 6 on the Arduino, which we'll enable as a secondary serial port by using Arduino's Software-Serial library (the Arduino Uno has a primary serial port that we'll use to send messages to the computer). You'll also need to attach the enable pin to ground. Figure 3-1 shows these connections.

Try It

This sketch reads in bytes similar to the Processing sketches shown in Chapter 2. Upload it to your Arduino, launch the Arduino Serial Monitor (Tools→Serial Monitor), and make sure the Serial Monitor is configured for 9600 bps. Next, bring an RFID tag within range of the reader, and you should see the tag ID appear in the Serial Monitor window.

```
/*
RFID Reader
*/

#include <SoftwareSerial.h> // Bring in the software serial library ❶

const int tagLength = 10;    // each tag ID contains 10 bytes ❷
const int startByte = 0x0A;  // Indicates start of a tag
const int endByte   = 0x0D;  // Indicates end of a tag

char tagID[tagLength + 1];   // array to hold the tag you read ❸

const int rxpin = 6; // Pin for receiving data from the RFID reader
const int txpin = 7; // Transmit pin; not used
SoftwareSerial rfidPort(rxpin, txpin); // create a Software Serial port ❹

void setup() {
  // begin serial communication with the computer
```

```
    Serial.begin(9600);

    // begin serial communication with the RFID module
    rfidPort.begin(2400);
  }

  void loop() {

    // read in and parse serial data:
    if (rfidPort.available() > 0) { // ❺

      if (readTag()) { // ❻
        Serial.println(tagID);
      }

    }

  }

  /*
   This method reads the tag, and puts its
   ID in the tagID
   */
  boolean readTag() {

    char thisChar = rfidPort.read(); // ❼
    if (thisChar == startByte) {    // ❽

      if (rfidPort.readBytesUntil(endByte, tagID, tagLength)) { // ❾
        return true;
      }

    }
    return false;
  }
```

Here's how the code works:

❶ This line imports a library called SoftwareSerial, which allows you to use any pair of digital pins as a serial port. It's not as robust as the built-in hardware serial port (pins 0 and 1), but it is well-behaved at low transmission rates, which is perfect here, since we're communicating with the RFID reader at 2400 bits per second.

❷ These lines of code declare several constants used throughout the sketch. The first, tagLength, is the number of characters in the RFID tags that the Parallax reader can process. The second, startByte, is the value of the character that the RFID reader sends when it's beginning to transmit a tag ID. And the last, endByte, is what the RFID reader uses to signal that it's done transmitting a tag ID.

❸ This line declares a buffer (`tagID`) that's big enough to hold the tag ID, along with an extra byte at the end to hold the character (zero) that terminates a character array. This way, when you use the `println` command to display the `tagID` later, the zero will signify to Arduino that it's reached the end of the string.

❹ This initializes the Software Serial session with pin 6 as the receiving pin, and 7 as the transmitting pin. Since you don't transmit anything to the RFID reader, you don't need to hook pin 7 up. In the `setup()` module, you'll see that the code calls `begin()` on both the built-in serial port (to talk to the computer and this port that you just initialized).

❺ This expression checks the Software Serial port to see if any messages are coming in from the RFID reader.

❻ If the previous expression evaluated to true, this line calls the `readTag()` function. If that returns true, the next line (`println()`) displays the tag to the built-in serial port, which makes it appear on the Serial Monitor's display.

❼ This line reads one character from the RFID reader, and stores it in `thisChar`.

❽ If `thisChar` is equal to the `startByte` indicator, the next line is run.

❾ The `readBytesUntil` function will read bytes from the RFID reader until it hits the `endByte` delimiter, and it stores the result into `tagID`. If it succeeds in reading `tagLength` (10) bytes from the RFID reader, this function returns true.

Here's the sort of output you'll see in the Serial Monitor as you bring different tags in range (since you have different tags than I do, you'll see different values there). Because the Parallax reader transmits a tag ID continuously while it is in range, you'll see the ID repeated as long as you hold it next to the reader:

```
04162F7CAC
04162F7CAC
0415EA09BE
0F02A684B1
0F02A684B1
0F02A684B1
```

Searching for a Specific Tag

With the same circuit, and just a few changes to the code, you can make the Arduino take action only when a certain tag comes within range. Before you try this, you'll need to run the sketch from the previous section, and copy

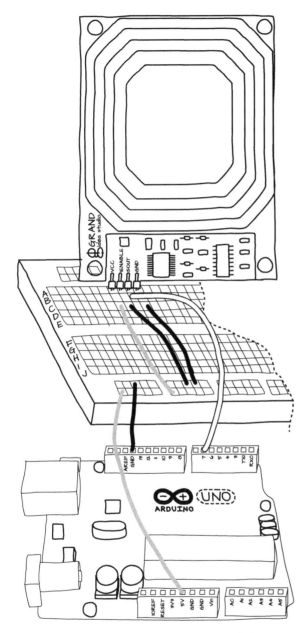

Figure 3-1. *Wiring the RFID reader to the Arduino*

the ID of the tags you want to match. For this example, I'll use `04162F7CAC`, but you will need to use a tag ID from your collection of tags.

Here's the modified sketch, which lights the Arduino's built-in LED when the right tag is brought within range of the reader:

```
/*
 RFID Reader
 */

#include <SoftwareSerial.h> // Bring in the software serial library

const int tagLength = 10;    // each tag ID contains 10 bytes
const int startByte = 0x0A;  // Indicates start of a tag
const int endByte   = 0x0D;  // Indicates end of a tag

char tagID[tagLength + 1];   // array to hold the tag you read

const int rxpin = 6; // Pin for receiving data from the RFID reader
const int txpin = 7; // Transmit pin; not used
SoftwareSerial rfidPort(rxpin, txpin); // create a Software Serial port

String matchingTag = "04162F7CAC"; // The tag to match ❶
const int ledPin    = 13; // The digital pin for the built-in LED ❷

void setup() {
  // begin serial communication with the computer
  Serial.begin(9600);

  // begin serial communication with the RFID module
  rfidPort.begin(2400);

  pinMode(ledPin, OUTPUT); // enable the pin for output ❸
}

void loop() {

  // read in and parse serial data:
  if (rfidPort.available() > 0) {

    if (readTag()) {
      Serial.println(tagID);
      if (matchingTag.equals(tagID)) { // ❹
        digitalWrite(ledPin, HIGH); // Turn on the LED ❺
      }
      else {
        digitalWrite(ledPin, LOW);  // Turn it off ❻
      }
    }

  }

}
```

```
/*
 This method reads the tag, and puts its
 ID in the tagID
 */
boolean readTag() {

  char thisChar = rfidPort.read();
  if (thisChar == startByte) {

    if (rfidPort.readBytesUntil(endByte, tagID, tagLength)) {
      return true;
    }

  }
  return false;
}
```

Here's what those new lines of code do:

❶ This is an Arduino **String** variable that holds the ID of the tag you're trying to match.

❷ This constant specifies which digital pin to use to light the LED. Digital pin 13 corresponds to the LED that's built into the Arduino board.

❸ This configures the pin so you can use it for output.

❹ This line checks to see if the tag ID you just read from the reader matches the one you seek.

❺ If so, it lights the LED by writing **HIGH** to it.

❻ If not, it turns it off by writing **LOW**.

These last few lines of code didn't add a lot, but they opened the door to a powerful new capability: taking action in the physical world based on the bytes you received from the RFID reader. In Chapter 4, you'll see how to go from turning on an LED to turning on an actual light (or any other electrically-powered device).

4/RFID Meets Home Automation

Between my officemate and me, we have dozens of devices drawing power in our office: two laptops, two monitors, four or five lamps, a few hard drives, a soldering iron, Ethernet hubs, speakers, and so forth. Even when we're not here, the room is drawing a lot of power. What devices are turned on at any given time depends largely on which of us is here, and what we're doing. This project is a system to reduce our power consumption, particularly when we're not there.

When either of us comes into the room, all we have to do is tap our key fobs on a reader mounted by the door, and the room turns on or off what we normally use. Each of us has a keyring with an RFID-tag key fob. The reader by the door reads the presence or absence of the tags.

The reader is connected to a microcontroller module that controls the AC power lines using a device called the PowerSwitch Tail, shown in Figure 4-1. Each of the various power strips is plugged into one of these. Depending on which tag is read, the microcontroller knows which power strip to turn on or off.

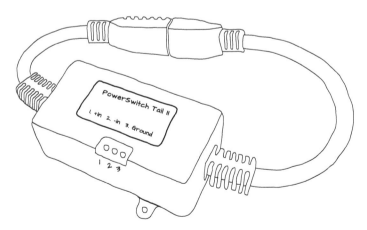

Figure 4-1. *The PowerSwitch Tail II*

Materials

Arduino

    The Arduino Uno (*http://www.makershed.com/product_p/mksp4
.htm.*) is a good model of Arduino to get started with, and it will work for
all the microncontroller-based projects in this book.

RFID reader

    Parallax's RFID Reader Module, available from Maker Shed as part of a
starter pack (*http://www.makershed.com/product_p/mkpx16.htm*) or
by itself (*http://www.makershed.com/product_p/mkpx2.htm*).

RFID tags

    The starter pack (*http://www.makershed.com/product_p/mkpx16
.htm*) includes several tags, and you can buy them
separately (*http://www.makershed.com/SearchResults.asp?Search=rfid*).

Breadboard

    You can use a half-size (*http://www.makershed.com/product_p/
mkkn2.htm*) or mini breadboard (*http://www.makershed.com/Mini
_Breadboards_p/mkkn1.htm*) to make connections between the reader
and the USB-to-TTL serial adaptor.

2 PowerSwitch Tails

    Each one can control 15 amps of current. They are available from Maker
Shed (*http://www.makershed.com/PowerSwitch_Tail_II_p/mkps01
.htm*).

The Circuit

The RFID module is connected to the microcontroller as in Chapter 3. Connect the microcontroller to the PowerSwitch Tails as shown in Figure 4-2:

First PowerSwitch Tail

- "1: +in" to Arduino digital pin 2
- "2: -in" to breadboard ground

Second PowerSwitch Tail

- "1: +in" to Arduino digital pin 3
- "2: -in" to breadboard ground

WARNING: This project controls high-voltage, high-current alternating current. Even though the PowerSwitch Tail is designed to make this safe, you should take extra care when wiring or rewiring the circuits for this project. Make sure everything is wired correctly and mounted securely *before* you plug either the Arduino or the PowerSwitch Tails in.

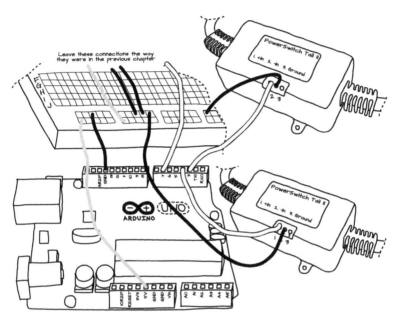

Figure 4-2. *Arduino connected to the PowerSwitch Tail II*

Try It

Run this sketch to test the PowerSwitch Tail:

```
/*
 PowerSwitch Tail test
 language: Wiring/Arduino

*/

const int switchOne = 2; // the first PowerSwitch Tail control pin ❶
const int switchTwo = 3; // the second PowerSwitch Tail control pin
```

```
void setup() {
  pinMode(switchOne, OUTPUT); // Configure the pins ❷
  pinMode(switchTwo, OUTPUT);

  digitalWrite(switchOne, LOW); // Make sure they are off ❸
  digitalWrite(switchTwo, LOW);
}

void loop() {

  // turn on first power strip, turn off the second ❹
  digitalWrite(switchOne, HIGH);
  digitalWrite(switchTwo, LOW);
  delay(2000);

  // turn on second power strip, turn off the first
  digitalWrite(switchTwo, HIGH);
  digitalWrite(switchOne, LOW);
  delay(2000);

}
```

Here's what happens in the sketch:

❶ First, the sketch initializes two variables, switchOne and switchTwo, to hold the pin numbers of the two PowerSwitch Tail units you have connected.

❷ In setup(), you configure the pins to be OUTPUTs.

❸ Next, we make sure the pins are off (this is the default, but it makes it clear that your sketch expects to start with everything off).

❹ The loop() function repeatedly turns the first switch on and the second off, then waits two seconds, and does the opposite.

The result is that each PowerSwitch Tail will turn on and off every two seconds. You probably don't want to leave this running for very long, unless you're trying to do endurance tests for light bulbs (or whatever you have plugged in).

Now that you've got control over your PowerSwitch Tails, you can combine the RFID and the PowerSwitch Tail code.

Switching Power with RFID

The following sketch will toggle the state of each PowerSwitch Tail when the corresponding RFID tag is held up to the reader. You can now use the RFID reader like a light (or power strip) switch:

```
/*
RFID-to-PowerSwitch Tail control
language: Wiring/Arduino

*/
#include <SoftwareSerial.h> // Bring in the software serial library ❶

const int tagLength = 10;    // each tag ID contains 10 bytes
const int startByte = 0x0A;  // Indicates start of a tag
const int endByte   = 0x0D;  // Indicates end of a tag

const int rxpin = 6; // Pin for receiving data from the RFID reader
const int txpin = 7; // Transmit pin; not used

String currentTag;    // String to hold the tag you're reading ❷

String tag[] = { "04162F7CAC", "0415EA09BE"}; // List of tags ❸
int numTags = 2;                              // Number in that list

// PowerSwitchTail unit pins and unit states: ❹
int numUnits    = 2;            // Two PowerSwitch Tails
int unit[]      = {2, 3};       // Pins 2 and 3
int unitState[] = {LOW, LOW}; // Both start in the off position

long lastRead;         // the time when we last read a tag
long timeOut = 1000; // required time between reads

SoftwareSerial rfidPort(rxpin, txpin); // create a Software Serial port

void setup() {

  lastRead = millis(); // Initalize to the sketch's start time ❺

  // begin serial communication with the computer
  Serial.begin(9600);

  // begin serial communication with the RFID module
  rfidPort.begin(2400);

  // Initialize all the PowerSwitchTail controller pins. ❻
  for (int thisUnit = 0; thisUnit < numUnits; thisUnit++) {
    pinMode(unit[thisUnit], OUTPUT); // Enable this pin for output
    digitalWrite(unit[thisUnit], unitState[thisUnit]);
  }
}

void loop() {
  // read in and parse serial data: ❼
  if (rfidPort.available()) {
    readByte();
  }
```

```
}

void readByte() {

  char thisChar = rfidPort.read(); // Read a character from the port ❽

  // depending on the byte's value,
  // take different actions:
  switch(thisChar) {
    // if the byte == startByte, you're at the beginning ❾
    // of a new tag:
  case startByte:
    currentTag = "";
    break;
  //if the byte == endByte, you're at the end of a tag: ❿
  case endByte:
    checkTags();
    break;
  // other bytes, if the current tag is less than   ⓫
  // 10 bytes, you're still reading it:
  default:
    if (currentTag.length() < 10) {
      currentTag += thisChar;
    }
  }
}

void checkTags() {

  // iterate over the list of tags: ⓬
  for (int thisTag = 0; thisTag < numTags; thisTag++) {

    // if the current tag matches the tag you're on: ⓭
    if (currentTag.equals(tag[thisTag])) {

      // Only flip a switch if the tag has been away for a while ⓮
      if (lastRead + timeOut < millis()) {

        // unit number starts at 1, but list position starts at 0:
        Serial.print("unit " + String(thisTag +1));

        // change the status of the corresponding unit: ⓯
        if (unitState[thisTag] == HIGH) {
          unitState[thisTag] = LOW;
          Serial.println(" turning OFF");
        }
        else {
          unitState[thisTag] = HIGH;
          Serial.println(" turning ON");
        }
```

```
        // Set the switch to the new state. ❶⓺
        digitalWrite(unit[thisTag], unitState[thisTag]);
      }
      lastRead = millis(); // mark the last time you got a good tag ❶⓻

    }
  }
}
```

❶ The top part of the sketch is essentially the same as what you saw in the example in Chapter 3.

❷ This string contains the tag that's currently being read. As the RFID reader collects each byte of the tag ID, it will add it to this string until it's full.

❸ This array contains a list of tags that the sketch will accept. This array should have the same number of elements as the array (`unit[]`) that you'll see later. That's because each element in this array corresponds to one PowerSwitch Tail unit. That means that the first RFID tag in the `tag[]` array will turn the first PowerSwitch Tail on and off, and the second RFID tag will turn the second PowerSwitch Tail on and off.

❹ Here's where you set up an array (`unit[]`) containing the Arduino pin number corresponding to a single PowerSwitch Tail. You also set up an array to track the current state (`HIGH` or `LOW`) of each unit.

❺ This variable is used to make sure you don't continuously toggle a switch on and off as long as you hold a tag in place. The Parallax reader continuously transmits a tag as long as you hold it in place. Later on, you'll see that the sketch requires that you keep the tag away for at least one second before it will let you toggle a switch.

❻ This loop goes over each pin in the `unit()` array and sets its initial state (`LOW`, or off).

❼ In the `loop()` function, you look for any serial activity from the RFID reader, and call `readByte()` if there is any.

❽ Here, we read one character from the RFID reader.

❾ This `switch` statement takes action based on what was read from the RFID reader. If it sees the start of tag byte, it initializes the `currentTag` string.

❿ If you reached the end of a tag, call `checkTags()`.

⓫ Otherwise, keep appending the character to the `currentTag` string, as long as you don't exceed the maximum tag length.

⓬ At the top of `checkTags()`, start a loop that iterates over the list of tags.

⓭ For each of the tags in the list, check to see if it matches the one you just read from the RFID reader.

⓮ This line checks the `lastRead` variable to make sure you've waited long enough before toggling a switch.

⓯ Here's where you toggle the switch's state: if it's on, set it to `LOW`, or off. If it's off, set it to `HIGH`, or on.

⓰ This line calls `digitalWrite()` to turn a PowerSwitch Tail on or off.

⓱ Finally, set `lastRead` to the current time (the time at which you last read a valid tag).

Note that the reader lacks the ability to read multiple tags if more than one tag is in the field. That's an important limitation. It means that you have to design the interaction so that the person using the system places only one tag at a time, then removes it before the second one is placed. In effect, it means that two people can't hold their key tags to the reader at the same time. In other words, users of the system need to take explicit action to make something happen. Presence isn't enough. That's why I mounted the reader vertically on a wall, so that tags wouldn't just lay on the antenna all the time.

5/Conclusion

Despite what some authors would have you think, RFID is pretty far from being a "big brother" technology. The "identity" of each RFID tag is just a number. What you do with that information is up to you. There is no single universal RFID protocol, nor are there readers that can read every tag out there. There's no secret database somewhere that holds all the RFID numbers and associates them with the things they're attached to, or worse, the people who own those things. To make RFID meaningful, you have to build your own database associating the tags you distribute with the objects to which you attach them, and you have to make sure that you're using readers that can read that particular brand of tag. The range of passive RFID systems is very short, and most of the time you have to be very close to a tag to read it. With most readers, you can only read one tag at a time; a cluster of tags together generally won't be read. If you're really concerned about the security of an RFID tag, you can always wrap it in a shield (*http://lessemf .com/*) of conductive fabric or aluminum foil, which will make it difficult or impossible to read. There is a lot you can do with RFID within those limitations, and the more you understand how it works, the more you can use it to your advantage.

The projects shown here barely scratch the surface of what you can do with RFID. Even staying within the limits of passive RFID tags and inexpensive readers, there is more you can do. For example, most tags are not just readable, but also writeable; you can store small amounts of data on them. To do this, you need a reading device that's capable of both reading from tags and writing to them, unlike the one shown here.

If you're interested in other passive RFID readers, several of them are covered in *Making Things Talk*, second edition. A few of my favorites are listed below:

SonMicro's SM130 series (http://www.sonmicro.com/)
> These are reader/writer modules for the 13.56 MHz ISO14443A Mifare standard tags. Mifare is a very popular tag standard, used by many transit systems around the world. SonMicro's modules give you the ability to both read to and write from these tags.

ID Innovations' readers (http://id-innovations.com/httpdocs/products .htm)

These cover a wide range of physical forms and RFID protocols. The company makes some super-compact ID-12 and ID-20 readers for 125KHz tags, and recently added ID-20MF Mifare read/write modules as well. It also makes long-range passive RFID readers, which claim to be able to read up to 1.2 meters.

Texas Instruments (http://www.ti.com/rfid/)

TI makes a wide range of both passive and active RFID readers, particularly in the 134.2kHz range that's popular with pet tagging systems.

Adafruit's PN532 RFID and NFC reader (https://www.adafruit.com/prod ucts/364)

This reader is particularly interesting to try. Compatible with the Mifare tags, it can also communicate with near field communications devices that are coming on the market now. NFC is an extension of RFID, a communications protocol for devices that are very close to each other or touching. Look for many interesting NFC applications in the near future, as cards, tags, and other small objects begin to acquire interesting behaviors.

About the Author

Tom Igoe likes playing with electronics, mechanics, and programming; making things that let people express themselves; and unusual clocks. He has written two books for makers, *Physical Computing* with Dan O'Sullivan and *Making Things Talk*, and is a contributor to *Make* magazine. He teaches at the Interactive Telecommunications Program at NYU. He is a cofounder of Arduino because he believes that open fabrication can change the world. He is a fan of women's flat-track roller derby and lives in Brooklyn with a cat named Noodles. He is currently realizing his dream of working with monkeys, and wants to visit Svalbard someday.

Get even more for your money.

Join the O'Reilly Community, and register the O'Reilly books you own. It's free, and you'll get:

- $4.99 ebook upgrade offer
- 40% upgrade offer on O'Reilly print books
- Membership discounts on books and events
- Free lifetime updates to ebooks and videos
- Multiple ebook formats, DRM FREE
- Participation in the O'Reilly community
- Newsletters
- Account management
- 100% Satisfaction Guarantee

Signing up is easy:

1. Go to: oreilly.com/go/register
2. Create an O'Reilly login.
3. Provide your address.
4. Register your books.

Note: English-language books only

To order books online:
oreilly.com/store

For questions about products or an order:
orders@oreilly.com

To sign up to get topic-specific email announcements and/or news about upcoming books, conferences, special offers, and new technologies:
elists@oreilly.com

For technical questions about book content:
booktech@oreilly.com

To submit new book proposals to our editors:
proposals@oreilly.com

O'Reilly books are available in multiple DRM-free ebook formats. For more information:
oreilly.com/ebooks

Spreading the knowledge of innovators   oreilly.com

Have it your way.